Welcome to
The Funny Farm

by

CAROLYN SANDERS

Welcome to our Homestead
First edition, published 2021

By Carolyn Sanders
Cover and interior design by Reprospace.com

Copyright © 2021, Carolyn Sanders

Hardbound ISBN-13: 978-1-952685-11-8

All rights reserved. No part of this book may be reproduced or transmitted in any form or by any means, electronic or mechanical, including photocopying, recording or by any information storage and retrieval system, without written permission from the author, except for the inclusion of brief quotations in a review.

Published by
www.KitsapPublishing.com

Dedication

I am dedicating this book in memory of my loving mother,
Sharon Lee Holt.

Introduction

In 1996, we settled in Poulsbo, Washington, and turned three acres of thick alder into a home and thriving little farm.

We cleared our property and planted a garden measuring 20 by 40 feet in the spring of 1997. It was the first time I had ever dug potatoes, pulled ears of corn from a stalk, and picked over a bushel of beans. The potatoes just kept coming in all shapes, sizes, and colors! My daughter, mother, and I were sitting in the dirt laughing with joy. We could not believe our harvest. Our hearts were full.

In 1998, we expanded our garden to 40 by 60 feet and sold our produce at our local Farmer's Market and local restaurants. The community embraced our farm's produce.

In 1999, we expanded our garden to 90 by 120 feet and added garden subscriptions, which created a consistent sale of produce. Our community asked if they could visit our farm and pick vegetables.

In 2000, we opened to the public on Wednesday and Sunday. Customers spent as much time as they liked selecting and harvesting their favorite veggies. They weighed their chosen produce in a little garden shed. By the honor system, customers paid $2.50 per pound. People happily came and went from our farm. Eventually, the community started calling our farm the "The Funny Farm." People would laugh as they watched the farm's antics.

As the children grew, our farm became a means to self-sufficiency. We turned inward, falling into a seasonal pattern that ebbed and flowed. It was hard work, but we planned, prepared, planted, tended, harvested, and put up our food. Some years we added pigs, chickens, geese, turkeys, ducks, or sheep; some years were more productive than other years. The years cycled based on our family needs.

Our pantry was full of jams, syrups, fruits, relishes, and pickled veggies; the freezer was full of veggies and meat. We filled our dry storage with squash, carrots, and Potatoes. Life was simple and felt natural.

I was inspired by how my children and friends' children loved working in our garden. I had no idea that some adults and children did not realize carrots grew in the ground or what an abundance of beans grew on one plant. I thought back to how I felt when I dug potatoes for the first time, and I wanted to share my learned farming knowledge to educate others they too could grow food.

It was this year, 2014, that I first wrote and shared this Little Funny Farm book. Instead of selling vegetables, I sold Veggie Pots and Salad Bowls. Folks that never gardened before purchased these beautiful containers for their patio. Space was no longer an excuse, and my community felt empowered.

I have learned in my 54 years that family and community, above all else, are the most valuable resources in my life, and I'll only have as much strength in life as I allow myself.

Go forward and plant vegetables.

You can do it!!

Winter, February

As winter ends,

the baby animals arrive.

The first babies to arrive in late February are the piglets. Just look at all his hair! A piglet is called a "Weaner Pig" from the time it is weaned from its mommy, up to 40 pounds. Did you know that pigs can smell food, like roots and worms from underground, and use their nose to dig up that food? We call such activities rooting. I like to watch pigs root; it looks like they are plowing a field.

Winter, March

In March, the bunnies are born.

They are called a litter of kits. The kits stay warm in a nest made of straw and rabbit hair their mommy pulled from her chest. Kits are born without fur and have pink delicate thin skin. These kits are five days old and have started to grow hair. Can you imagine how velvety soft, and warm these kits feel? Can you count the number of kits that lay sleeping in their nest?

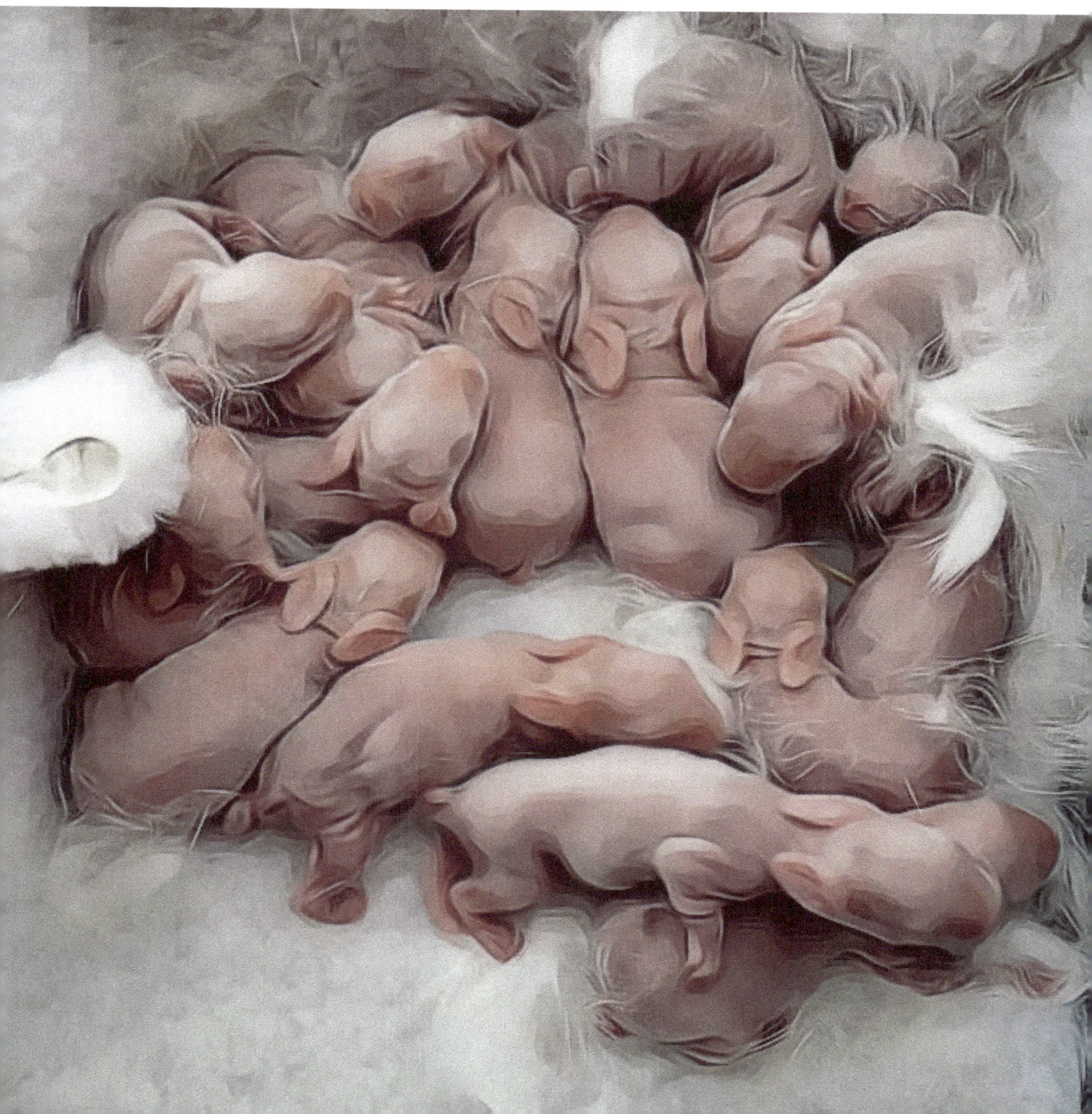

Spring, March & April

A female sheep is called an ewe.

When their babies are born, it is called lambing. Ewes typically have twins. On our farm, lambing happens in March and April. A ewe carries her lamb for 152 days. Our ewes are excellent mothers and lambs on their own. Isn't this newly born lamb a miracle? When this lamb turns one year old, it will be a sheep. A group of sheep is called a flock.

Spring, April

A mommy chicken is called a Brood Hen.

A brood hen sits on her nest of eggs for 21 days, waiting for her eggs to hatch. Her eggs will start to rock, and you can hear a faint peep from inside the egg. Soon the chick will begin pecking at the shell from the inside of the egg. The chick uses a special "egg tooth" at the tip of its beak to break out of its shell. The egg tooth falls off the tip of the beak after a few days.

Spring, April

The bunnies are now five weeks old and independent.

They have grown into busy hoppers bouncing about in their pen. The girls are called does and the boys are called bucks. Do you know why rabbits need to chew? They chew because their teeth never stop growing. Bunnies are super soft and cuddly. They make nice quiet pets and are easy to keep clean.

Spring, April

It is April and our barn cat, Momma Kitty, has had her litter of kittens.

It's April, and our barn cat, Momma Kitty, has had her litter of kittens. Kittens are born with their eyes and ears sealed shut. They can't hear or see, but they can smell, taste and feel. Their eyes and ears gradually open between two and three weeks of age. Then it is time to explore. Momma Kitty will watch over them, keeping them safe and warm for five to six weeks. Then she will teach them to hunt mice.

Spring, May

It is now May.
The weaner pigs are now grown enough to be called "Grower Pigs."

A grower pig weighs between 40 and 150 pounds. They live happily in their pig playpen. They run about snorting, grunting, barking, and jumping. Did you know an adult pig can run at speeds up to 11 miles per hour? Wow, that's pretty incredible. Rob enjoys feeding and caring for the pigs. Can you see their curly tails? Did you know when pigs are happy, they wag their tails like a dog?

Spring, May

It is still May and it is time to plant our vegetable garden.

This might not look like much right now-but just watch. It all starts with the power of a seed. Growing a garden is hard work but very simple. Just put a seed in the ground. Water and tend to that one seed and it will grow into a strong vegetable plant which will provide you with food. Many seeds will grow a crop.

Spring, May

These seed potatoes are ready to be planted.

Did you know that potatoes have eyes? Can you see the buds growing from the potatoes? The buds are the eyes. One potato has three to ten eyes and each eye creates a potato plant yielding five to seven potatoes. We cut the potato into sections and each section has one eye. Then we plant the sections. Can you find the cut section with an eye? Good job! Yes, that is a purple heirloom potato called an "All Blue." Delicious!

Spring, June
The raspberries are ready to be picked

In early spring, the raspberry canes flower, attracting hundreds of busy bees. These bees help pollinate the raspberry blooms, garden, and fruit trees. This process makes our raspberries an essential pollinator for our farm.

Raspberries are a true aggregate fruit, which are oodles of juicy balls called drupelets. Each drupelet contains a seed. As the raspberry matures, the drupelets form a tight cluster. When ripe, the raspberry slides off the receptacle, and the cluster of juicy balls stay together, forming the raspberry.

If a raspberry isn't picked and falls to the ground, the seed inside the drupelet will germinate, and a new raspberry cane will grow next year.

Late Spring in June

Ta-Da!! Here I am!

Can you point to the little wing emerging from the shell? The chick's feathers are moist and will dry in a matter of minutes. This chick is well on its way to a healthy, vigorous life.

Late Spring in June

Can you see this chicks foot?

Notice the tiny claws at the end of its toes.

Late Spring in June

This mommy Muscovy hen sat on her clutch of eggs for 35-days.

She is an excellent mother that will protect her chicks. A Muscovy hen will lay over 100 eggs a year and hatch up to four clutches. Sometimes hens will collectively lay a clutch of eggs and share sitting on their nest. They will also raise their chicks together. Muscovy are quiet ducks because they don't quack; they thrill. A thrill is a quiet, breathy coo. Muscovy are also land ducks, although they still like to go for a swim in the pool.

Summer, Early July

Our sheep does not have hair they have wool, and wool is considered a fiber.

When the fibers cling together, it produces felt. In early summer, the sheep get sheared. Shearing is when the wool is shaved off. Just like a haircut. The wool for one sheep, when shaved, is called a fleece. Such a task requires a sheep shearer. Look at our sheep shearer, shear our sheep. Would you like to do this job?

Summer, Early July

We have made up silly games to pass time.

Because we all work together as a family on our farm caring for the animals and growing our food in the garden, we have fun. We have made up silly games to pass time. My favorite is the game, "name that bird." We don't really know their names so we make up a silly name according to the sound it makes. My favorite bird sound is the "laughing bird." What is one of your favorite silly games?

Summer, July

We each have our favorite vegetables to grow.

Rob's are shelling peas and zucchini. McCanna's is snow peas. Dylan's are corn and beans. Mine are tomatoes and carrots. What is your favorite vegetable? Just look at all of these yummy vegetables growing. It won't be long now before we will be eating them. Did you eat a vegetable with your dinner last night?

Summer, July

Surprise!

One morning as we were tending to our chickens, three peacock brothers walked out of the blackberry bushes and decided to live with us. Only males are called peacocks, females are peahens, chicks are peachicks, collectively they are called peafowl, and they are colorful pheasants. A flock of peafowls is called a party. Peafowls are ground feeders dining on insects, plants, and small creatures. Their tail is more that 60% of their body length and called a train. They are the world's largest flying bird. In the evening they take flight to roost in tall evergreens. Why do you suppose this party of peafowl chose to come live with our family?

Summer, July

Meet Uhaul: He is a very special members of our farm family.

Horses are big animals and require a lot of daily care. McCanna takes care of the horses on our farm. It is her daily chore to feed and fill their water twice a day. One horse eats up to 15 pounds of hay and drinks up to 35 gallons of water a day. Wow! McCanna is careful to hold her hand flat when giving them treats, so they don't mistake her fingers for carrots. Did you know that horses also have to eat their vitamins every day? Just like you! Vitamins help them grow strong and stay healthy.

Summer, July

Ireland

Here is one of Momma Kitty's kittens that has grown into a very nice young cat. We call her Ireland. She is a very sweet, fun, fury, and entertaining friend. We really enjoy her and she makes us laugh.

Summer, August

This boy bunny has matured to be called a "Buck".

This is a buck rabbit that is now full grown. He weighs 12 pounds and likes to eat fresh grass. Didn't he grow up fast?

A buck is a fully grown male rabbit. Didn't he grow fast? Since he enjoys bouncing about searching for his favorite garden greens, we named him Mr. Bouncer. Mr. Bouncer's favorite greens to nibble are clover, dandelions, and grass. Rabbits manure is the perfect mix of nitrogen, phosphorous, and potassium. Rabbits help us keep our soil healthy.

Summer, August

Now it is August.
Our garden is thriving.

Our crop is almost ready to harvest. Do you know what harvest means? Harvest is the term used when it is time to gather a crop. This is a very happy and exciting time. We usually cook a big dinner, using our vegetables, in celebration. Can you imagine the smell of fresh hot pumpkin pie? Harvesting is time when we reap the rewards of our labor.

Summer, August

Carrots are a root vegetable because the orange part grows in the dirt and this is the part we eat.

Pulling carrots out of the ground is always exciting because the size of the carrot is a surprise. Just like people, carrots come in all shapes and sizes. Do you see the curly carrot? I wonder how that happened. Do you have any ideas? That is a real mystery. The farm animals love to eat the green carrot tops.

Summer, August

These are tasty yellow zucchini.

Did you know that zucchini are in the squash family which makes them a fruit? All squashes grow from a pollinated flower and have seeds. These two things make squash fruit. The flowers are edible and taste just like a zucchini. Since zucchini are a fruit, why do we call them a vegetable? Strange...my best guess is because we eat them as a vegetable.

Summer, August

Yes, these are Purple Pole Beans.

Cool! A pole bean plant is typically seven to eight feet in length. How tall are you? One plant will produce approximately six pounds of beans. A bean, like the squash, grows from a pollinated flower and has seeds, so does that make them fruit? The answer is "no." The flesh is a vegetable and contains the seed. Next time you eat a bean, split the bean open and see the seeds. Vegetables are very fascinating.

Summer, August

These are heirloom tomatoes.

An heirloom tomato is an old traditional variety that is still being planted and maintained by today's farmers and gardeners. An heirloom plant will re-seed itself naturally. Re-seed means that the seed will grow a plant in the soil where it fell, without any human assistance. Amazing! These tomatoes are sweet and yummy.

Summer, Late August

We have two new friends living with us.

The grey goose is an "African". The white goose is an "Embden". These amusing geese spend their day waddling about honking to each other. I enjoy imagining what they must be saying to each other. They also love swimming in their pool! We haven't given these geese names. Could you help us? What names would you give them?

Late Summer, September

Corn is also known as maize.

Did you know that corn is a type of grass and has ears? Wow! Do you think corn is a fruit, vegetable or grain? It wind-pollinates like some fruit, we eat corn as a vegetable, but when dried and ground for flour it is a grain. Interesting! Each yellow piece of corn is called a kernel. Each kernel is a seed. If you dry an ear of corn, you can plant those seeds next year. Corn is truly amazing!

Summer, September

Here I am with Grandpa in our thriving garden.

These super fun sunflowers attract bees and butterflies, which are needed to pollinate the berries, vegetables, and fruit trees. When the sunflower dries out, we harvest the plumpest sunflower seeds to plant next year and feed the rest to the chickens over the winter months. Sunflower seeds add valuable oil to our chickens' diet.

Early Fall in September

It is Early Fall and Grandma has come to visit and help harvest beans

She loves harvesting all of the vegetables. Her favorites are beans and potatoes. Look how happy she is. There is a surprise behind every leaf. Harvesting is a lot of fun for the whole family.

Fall, October

Here is a sample of the vegetables we harvested.

Can you name some of them? Do you see your favorite? Since we worked together, we supplied our family with vegetables for eight months. WOW!! As we eat our vegetables over the winter, we harvest the seeds, dry them, and then place them in a labeled paper envelope for planting next year's garden. Maybe you can try this in your home? Then you can reap the rewards of your labor.

Fall, October

The grower pig is now a Finisher Pig and weighs between 150-220 pounds.

After 220 pounds our pigs are ready to butcher. A pig provides meat such as pork chops, bacon and ham. Pigs are very fun to raise. The next best thing to a watching a happy pig is watching a happy dog play fetch. We feel blessed to provide a happy life for our piggies and healthy food for our family.

Fall, October

Our baby chicks are now six months old and are starting to lay their own eggs.

Can you tell which one is the rooster? His name is Carl. Can you count how many chickens there are? A hen will productively lay eggs for two years. A hen also needs 12 hours of light to stimulate and maintain egg production. Depending on the breed of chicken, over a five year lifespan, a hen will lay 160 to 800 eggs..

Fall, Late October

It is time to bring hay in for winter.

This is food for the horses. It is a very heavy job; the whole family and sometime friends come to help us as we often help our friend with their hay. We call it "hay season." One horse eats 20 pounds of hay a day; that equals 7,300 pounds of hay per year per horse. That is a lot of hay! Hay season brings the closing of the year to our small farm. Everything will start again come February.

In Memory of this Summer, see you next year

Thank you for visiting our Funny Farm!

We hope you had fun and learned something new.
You're welcome back any time.

www.ingramcontent.com/pod-product-compliance
Lightning Source LLC
Chambersburg PA
CBHW041218240426
43661CB00012B/1076